Practical Laboratory Skills Training Guides
Gas Chromatography

Practical Laboratory Skills Training Guides

Coordinating Author: Elizabeth Prichard, *LGC, Teddington, UK*

Series titles:
Gas Chromatography
by Brian Stuart, *LGC, Teddington, UK*

High Performance Liquid Chromatography
by Win Fung Ho and Brian Stuart, *LGC, Teddington, UK*

Measurement of Mass
by Richard Lawn, *LGC, Teddington, UK*

Measurement of pH
by Richard Lawn, *LGC, Teddington, UK*

Measurement of Volume
by Richard Lawn, *LGC, Teddington, UK*

Also available:
Practical Laboratory Skills CD-ROMs

For further information please contact:
Sales and Customer Care, Royal Society of Chemistry, Thomas Graham House,
Science Park, Milton Road, Cambridge CB4 0WF
Telephone: +44 (0) 1223 432360
Fax: +44 (0) 1223 426017
Email: sales@rsc.org

Practical Laboratory Skills Training Guides

Gas Chromatography

Brian Stuart
LGC, Teddington, UK

Coordinating Author
Elizabeth Prichard
LGC, Teddington, UK

Setting standards
in analytical science

VALID ANALYTICAL MEASUREMENT

A catalogue record for this book is available from the British Library

ISBN 0-85404-478-7

© LGC (Teddington) Limited, 2003

Published for the LGC
by the Royal Society of Chemistry,
Thomas Graham House, Science Park, Milton Road, Cambridge CB4 0WF, UK
Registered Charity Number 207890

For further information see the RSC web site at www.rsc.org

Typeset by Land & Unwin (Data Sciences) Ltd, Bugbrooke, Northants
Printed by Athenaeum Press Ltd, Gateshead, Tyne and Wear

Preface

Production of this set of five Training Guides and CD-ROMs was supported under contract with the Department of Trade and Industry as part of the National Measurement System Valid Analytical Measurement (VAM) programme.

The guides were written by staff at LGC in collaboration with members of the SOCSA Analytical Network Group whose assistance is gratefully acknowledged. They include liquid and gas chromatography, the measurement of mass, volume and pH.

Training has formed an essential part of the VAM programme since its inception in 1988. Many training courses on topics aimed at improving the quality of measurements have been developed. However, in working with groups of analytical scientists it has become clear that the basic skills required in an analytical laboratory are not covered on courses or readily available in paper format.

These guides are aimed at filling this gap and are aimed at those working at the bench. For each topic they include a limited amount of theory to explain the essential features but the main emphasis is on what to do to ensure reliable results. They contain references to further reading for those who wish to study the topics in more depth.

To help laboratory managers assess the competence of the trainee there are a limited number of exercises suggested. The chromatography modules also have a trouble shooting section.

The CD-ROMs cover Practical Laboratory Skills and have links to websites where more information may be obtained.

Contents

Gas Chromatography

1 Basic Theory

1.1 Overview

Today, gas chromatography is the most widely used technique in analytical chemistry – a position it has held for over three decades. The popularity and applicability of the technique is principally due to its unchallenged resolving power for closely related volatile compounds and because of the high sensitivity and selectivity offered by many of the detector systems. The technique is very accurate and precise when used in a routine laboratory.

1.1.1 Principle

The sample is normally introduced as a vapour on to the chromatographic column. On the column, the solubility of each component in the gas phase is dependent on its vapour pressure, which is in turn a function of the column temperature and the affinity between the compound and the stationary phase. Differences in vapour pressure cause the molecules of each component to partition between the mobile gas phase and the stationary phase. In fact, as the molecules are continually moving rapidly between the two phases, it is the difference in residence time in each phase that affects the partition. Every time a molecule enters the gas phase it is swept towards the detector by the carrier gas flow. Consequently, compounds having different physical and chemical properties will arrive at the detector at different times. The stationary phase can be a solid or a liquid coating an inert solid support, this gives rise to two forms of gas chromatography; gas–solid (GSC) and gas–liquid chromatography (GLC) respectively. Figure 1 shows the basic components of a gas chromatograph.

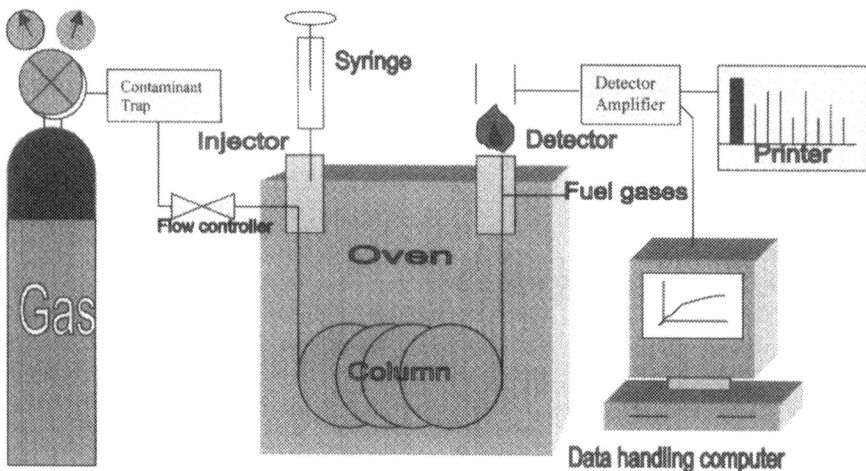

Figure 1 *Gas chromatograph*

1.2 Key Parameters

1.2.1 Retention Factor (k)

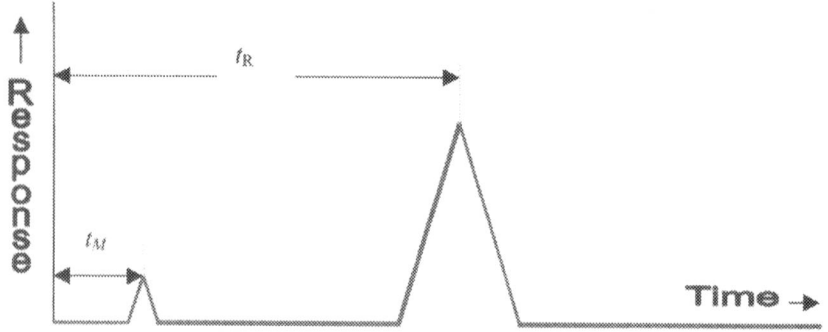

Figure 2 *Retention factor*

The retention factor (k) (see Figure 2) describes the ability of the stationary phase to retain a solute (Equation 1).

$$k = \frac{t_R - t_M}{t_M} \tag{1}$$

t_R retention time of the solute
t_M dead time (approximate time required for the mobile phase to pass through the column)

The longer the solute spends in the stationary phase the more likely it is to be separated from components of similar volatility. In gas chromatography, the retention factor can be altered by changing the stationary phase or the temperature of the column.

1.2.2 Selectivity Factor (α)

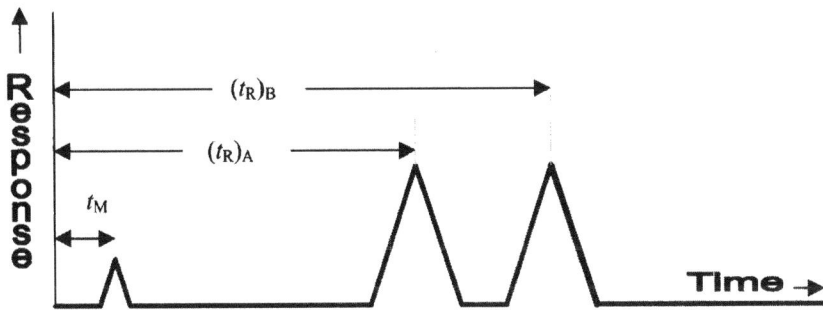

Figure 3 *Selectivity factor*

The selectivity factor (α) for two components provides a measure of how well they will separate on a particular column. Referring to Figure 3, the selectivity factor (α) is defined in Equation 2.

$$\alpha = \frac{(t_R)_B - t_M}{(t_R)_A - t_M} \tag{2}$$

$(t_R)_B$ retention time of component B which is more strongly retained
$(t_R)_A$ retention time of component A which is less strongly retained
t_M dead time

- When α = 1 then theoretically the two components cannot be separated on the column tested.
- When α > 1 then theoretically they can be resolved but this will depend on the resolution or performance of the column.

1.2.3 Column Resolution (R)

The resolution (R) of a column provides a quantitative measure of its ability to separate two components within a mixture.

Referring to Figure 4, the resolution is defined as shown in Equation 3.

$$R = \frac{2[(t_R)_B - (t_R)_A]}{W_A + W_B} \tag{3}$$

$(t_R)_B$ retention time of component B which is more strongly retained

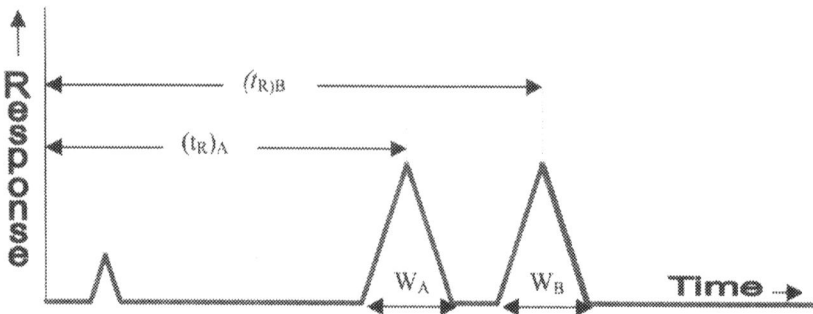

Figure 4 *Column resolution*

$(t_R)_A$ retention time of component A which is less strongly retained
W_A width of peak for compound A at the baseline
W_B width of peak for compound B at the baseline

- If R is less than 1, the component peaks are overlapping.
- If R is equal to or greater than 1, this indicates good separation.

1.3 Quantitative Measures of Column Efficiency

In Figure 2, Figure 3 and Figure 4 the peaks have been exaggerated to illustrate that the solute elutes from the chromatographic column as a band of material. This occurs because each molecule of the solute has a unique diffusion path through the liquid and gas phases. Consequently the molecules elute from the column at slightly different retention times. The effect is not only influenced by the chemical inter-action of solute and stationary phase but it is also highly dependent on the physical dimensions of the column, *i.e.* length, internal diameter or particle size, and coating thickness of stationary phase. This spreading (banding) effect increases in proportion to the length of time the solute molecules stay on the column; this explains why the later eluting peaks in a chromatogram are broader and shorter than the earlier peaks.

It is clear from this discussion that the ability of a column to separate similarly eluting components is related to how the peak or solute band spreads as it passes along the column, *i.e.* the efficiency of the column. Analysts use the plate model in chromatography to measure column efficiency; the higher the number of theoretical plates the higher the efficiency of the column.

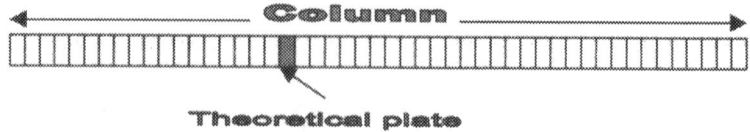

Figure 5 *Column as a series of theoretical plates*

This compares the chromatographic process with distillation and quantifies the efficiency of the column in terms of the number of theoretical plates. This comparison is much easier to understand for gas chromatography than other forms of chromato-graphy. Figure 5 shows the column as a series of theoretical plates.

Under 'simulated distillation' conditions – for example the hydrocarbon components of an oil elute from the chromatographic column in the same order as their boiling points – the components appear to boil off the chromatographic column as its temperature is increased in the same way as from a distillation column.

1.3.1 Number of Theoretical Plates (N)

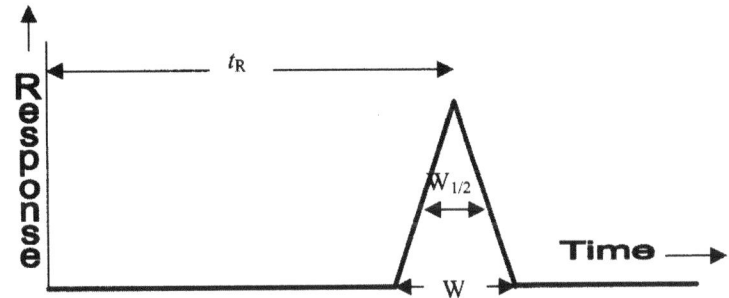

Figure 6 *Measurement of column efficiency*

t_R *retention time*
$W_{1/2}$ *width of the peak at 50% of the peak height*
W   *width of the peak at the baseline*

From Figure 6 the number of theoretical plates (*N*) is given by Equation 4.

$$N = 5.54 \times (t_R/W_{1/2})^2 = 16 \times (t_R/W)^2 \qquad (4)$$

where 5.54 and 16 are constants

Columns are available in different lengths (*L*) and so it is normal to express the efficiency of the column as the number of theoretical plates per metre. Alternatively the plate height (*H*) or height equivalent to a theoretical plate (HETP) is commonly used (Equation 5).

$$HETP = L / N \qquad (5)$$

Note that the value of HETP will be a minimum when the best efficiency for the column is achieved. The value of *N* and HETP will change as a column becomes older and less efficient. The analyst can use these parameters as a quantitative measure to determine when a column is no longer 'fit-for-purpose' and when it needs to be changed.

2 Carrier Gas

The carrier gas constitutes the mobile phase in the gas chromatographic system. During the chromatographic process component molecules from the sample are continually interchanging between the stationary and mobile phases (Figure 7). Every time these molecules enter the gas phase they are swept towards the detector by the flow of carrier gas.

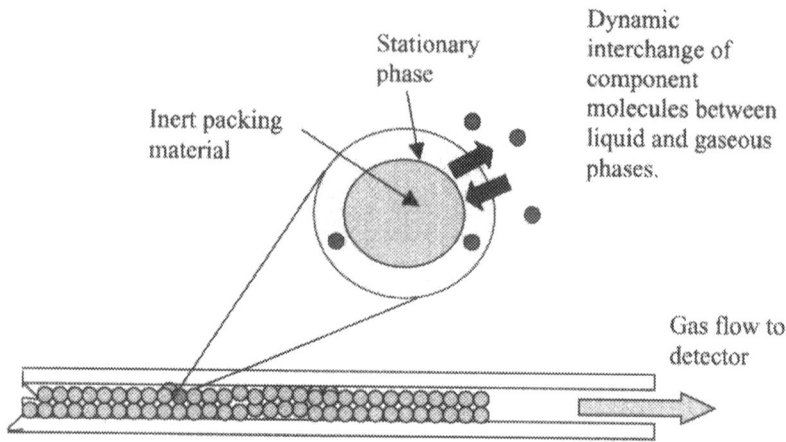

Figure 7 *Chromatographic processes on a packed column*

Consequently the carrier gas flow is an important variable in controlling the retention times of the sample components, the overall analysis, and the sample turn around times – *doubling the carrier gas flow halves the retention times of all the components on the column.*

Once a new chromatographic method is set up the carrier gas flow needs to be optimised to obtain the best chromatographic resolution from a particular column. The flow rate that provides the greatest resolution (HETP minimum) can be determined by simple experiment or by consulting a prepared van-Deemter plot.

The van-Deemter plots for a capillary column using three different gases are shown in Figure 8. The plots show that the curve for nitrogen is highly peaked and that the best resolution for this gas occurs over a narrow range of low flow rates (average linear velocity for capillary columns). In contrast the plots for helium and hydrogen are very flat and the best resolution is seen to occur at much higher flow rates. Higher flow rates mean much faster analyses and so there is a significant performance gain on using these gases in preference to nitrogen. Hydrogen often is regarded as the 'ideal' carrier gas, however its use has safety implications so most chromatographers use helium as it is non-flammable and therefore safer. For low resolution, *i.e.* packed column work, the van-Deemter curves for these gases are similar in shape and there is little advantage in using helium or hydrogen. In these cases the choice of carrier gas is governed by the availability, safety and cost of the gas, or its compatibility with the detector that is to be used.

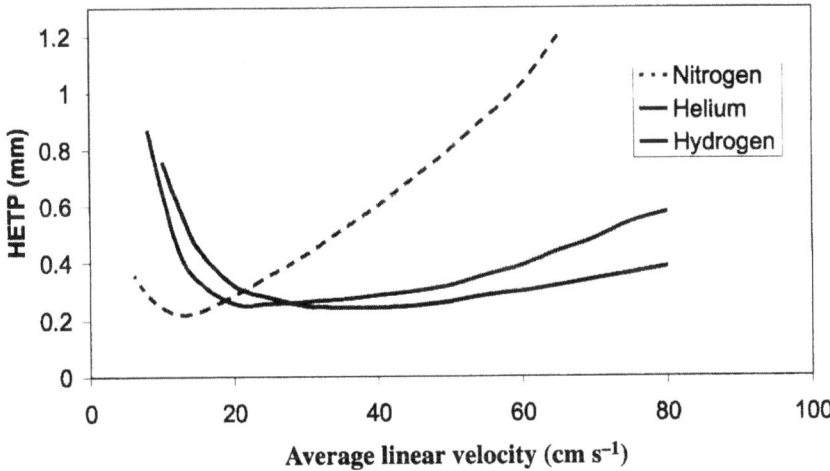

Figure 8 *Column HETP v average linear velocity for three gases*

2.1 Technical Requirements

All the gases used in GC analysis need to be of very high purity and steps must be taken to remove trace levels of moisture, oxygen and hydrocarbons from the gas before it enters the instrument. Oxygen is a particularly important contaminant to remove because it promotes column degradation and can also cause problems with the more sophisticated detectors such as electron capture and mass spectrometer detectors. Contaminants are removed by passing the gas through cartridges of absorbent. These are often 'self indicating' so the analyst can easily check if the absorbent bed is exhausted.

Gases are supplied in cylinders at high pressures so, unlike HPLC, there is no requirement for pumps to facilitate the movement of the mobile phase through the instrument. Cylinder pressures are attenuated and regulated precisely to ensure a constant pressure at the front end of the column. Mass flow controllers ensure that the pre-set flow-rate of gas is always constant during the analysis and is independent of the column oven temperature.

2.2 Tips

- The provision of gases for gas chromatographic analyses is the responsibility of the user. Always use high purity carrier gases, typically with purity of 99.995% or better, and note the specific recommendations of manufacturers and suppliers.
- Follow the guidance of manufacturers and suppliers to select the most effective system for removing trace contaminants from gases. Check regularly that gas-trap components are working effectively and replace the absorbent cartridges whenever necessary.
- Prevent the ingress of air into the instrument gases by leak-testing all connectors and joints in the gas delivery system. This is not only important when the

instrument is first set up, but also when the column or detectors have been changed.

● Always check that the reserves of gas are sufficient to allow the instrument to be left running overnight and at weekends.

3 Injection and Sampling Methods

3.1 Injection Methods

The function of the injector is to introduce a representative portion of the sample as a narrow band on to the chromatographic column. Because most samples that are analysed by gas chromatography are liquids, an essential feature of the injection stage is that the sample and solvent are vaporised prior to reaching the column. To accomplish this, injectors are equipped with dedicated heaters that control the temperature of the injector zone to a pre-set value.

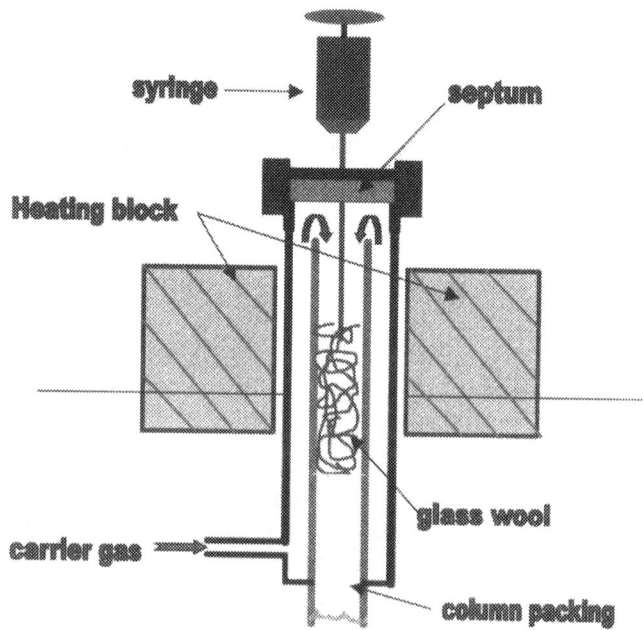

Figure 9 *Schematic-injector for packed column*

One of the earliest and simplest injector designs, developed for packed column systems, is shown in Figure 9. The sample is injected through the septum directly on to a glass wool plug at the top of the column. This region of the column is at high temperature and on injection the sample is vaporised and the flow of carrier gas ensures the vapour is pushed downwards in the direction of the analytical column.

3.1.1 'On-column' Injection

The packed column injector design is an example of 'on-column' injection where the whole of the sample passes through the chromatographic column. 'On-column' injection is also possible with capillary columns but here the sample is normally injected on to a cold column. The column temperature program is then activated to facilitate the separation and elution of the sample components. This cold 'on-column' technique is most suitable for samples that contain labile components or components having a wide range of volatility. To avoid overloading and droplet formation at the front of the capillary column, the injection volume is normally kept quite small, in the region of ≤ 0.3 µL. Consequently, detection limits are relatively high and the technique is not readily suited to trace analysis. Highly accurate quantitation is possible with this injection technique.

3.1.2 Split/Splitless Injector

This is a common injector system for use with capillary columns and has two modes of operation: split and splitless.

Split mode – Only a small fraction of the sample is introduced on to the column, the remainder is directed to waste according to a pre-set 'split ratio'. A high split ratio ensures that the column is not overloaded with sample at the beginning of the analysis. Some sample discrimination can occur to the detriment of the higher boiling components of the sample. For accurate quantitation the solvent and injection conditions must be carefully selected and controlled accurately.

Splitless mode – The sample is injected into a hot injector but the column is initially held at a temperature 15 °C to 30 °C below the boiling point of the solvent. Solvent and thermal focusing techniques can be used to concentrate the sample into a narrow band on the front of the column. This technique is useful for the determination of involatile analytes in low boiling solvents at very low concentrations. Commonly, a sample volume of 1 µL to 2 µL is injected into the heated injector.

3.1.3 Programmed Temperature Vaporiser

This injector has 'universal' features in that it is possible to operate it in hot or cold, split or splitless, total sample introduction or solvent elimination modes. The advantages and disadvantages of each mode have already been mentioned but the use of a temperature-programmed facility helps maintain the optimum conditions for the successful sample injection, *e.g.* for thermal focusing. The injector temperature can be programmed to follow column oven temperature changes. In the on-column mode the retention times of components can be controlled to ensure precision that provides a coefficient of variation between 0.1% to 0.5%. This is highly desirable, particularly when components are identified by their retention times alone.

3.2 Sampling Methods

3.2.1 Headspace Sampling

When a solid or liquid sample is placed in a closed container an equilibrium is formed between the volatile components in the bulk of the material and the headspace above the material. In the *static* headspace method small samples are taken from the sample headspace using a gas syringe and these are injected directly on to the gas chromatography column. This technique is quick and simple and, because samples are free of a matrix, the analysis is fairly straightforward. The main disadvantage of the technique is that for low sample concentrations, *e.g.* analytes of low volatility, the method is very insensitive. The situation can be improved using a *dynamic* headspace method, an example of which is the purge and trap technique.

3.2.2 Purge and Trap

In the purge and trap system, a stream of inert gas is used to strip components from an aqueous sample. The components are trapped and concentrated on an adsorbent trap and any water vapour passes through the trap unaffected. The trap is then thermally desorbed and the components are carried on to the column in a stream of carrier gas. Because the water matrix is almost entirely removed by the process, the sample is greatly concentrated and the technique allows the detection of parts per billion levels of volatiles in aqueous samples.

3.2.3 Thermal Desorption

In this technique trace contaminants in air are trapped on adsorbent tubes. The tubes are then thermally desorbed into a stream of helium carrier gas and passed to the column. To help focus the samples a low volume cryogenic trap is often employed prior to gas chromatographic analysis, *i.e.* two stage desorption. Because the traps do not adsorb the permanent gases (*e.g.* nitrogen, oxygen, *etc.*) in the original sample the method involves significant concentration of the analytes and is capable of detecting trace concentrations of contaminants in air.

3.2.4 Pyrolysis

Many large molecules, particularly polymers, cannot be analysed directly by gas chromatography due to their low volatility. In the pyrolysis method thermal energy is used to decompose the materials into simpler, lower molecular weight, fragments which can be analysed by gas chromatography. A number of different thermal sources are in use such as 'ribbon-type', 'curie point' and 'laser pyrolysers'. The technique is suitable not only to identify or determine the initial composition of the polymer but also to provide an insight into the thermal stability and thermal decomposition processes relevant to the material.

3.2.5 Derivatisation

For involatile compounds it is necessary to produce a volatile derivative. Derivatisation may be employed in chromatography to change the chromatographic behaviour or to improve detectability, and involves chemical modification of functional groups in the compound. Specific improvements in volatility, hydrolytic and thermal stability, peak shape, chromatographic resolution, detection selectivity and sensitivity can be achieved if a suitable derivatisation reaction is used. Typical functional groups that can be readily derivatised are; alcohols, phenols, carboxylic acids, amines, amides and thiols. Typical derivatising agents are trimethyl silyl, benzyl, benzoyl and phenoxyacyl halides. The presence of halogen atoms in the derivative ensures enhanced sensitivity and low detection limits using the electron capture detector.

3.3 Technical Requirements

It is suggested that operators should consult the manufacturer's manual to familiarise themselves with injector characteristics. Run standards and check for unusual or distorted peaks that might indicate the injector is not functioning correctly.

3.4 Tips

- Distorted or split peaks are an indication that too large a sample has been injected.
- Too low an injector temperature can lead to tailing peaks.
- Too high an injector temperature can lead to component degradation.
- Broad peaks at the beginning of a chromatogram are an indication that the solute has not been focused tightly into a band by the sampling and injection techniques. Check that the parameters associated with these techniques have been optimised and set correctly.
- Injector contamination can lead to memory effects – extra peaks in the chromatogram. Clean or replace the injector liner regularly.

4 The Chromatographic Column

The chromatographic column is the heart of the gas chromatograph, providing the necessary separation of the sample components before they are detected, identified or quantified. Early commercial columns consisted of long glass or metal tubes, $1/4$ inch diameter, packed with minerals, brick dust, molecular sieves (for GSC) or inert particles coated with gums, waxes and polymers (for GLC). The use of packed columns has declined as the higher performance open tubular capillary columns have become available.

Performance criteria demand that open tubular columns should be long and of narrow bore with an even distribution of stationary phase. Modern methods of extrusion, similar to those used in the manufacture of fibre optics, together with advanced coating techniques, where the stationary phase is coated or bonded to the inner walls of the tubing, have revolutionised the production of open tubular columns. Today, off-the-shelf columns with guaranteed performance, and customised specialist columns, are readily available.

There are a variety of types of open tubular columns, the most important are:

- WCOT – wall coated open tubular column
- SCOT – support coated open tubular column
- PLOT – porous layer open tubular column

4.1 Stationary Phases

The solubility of a compound in the stationary phase depends on the various types of intermolecular forces involved. The general rule is that like dissolves like, so that a phase suitable for the separation of a semi-polar compound is likely to be semi-polar.

The main types of stationary phases in use include:

- Solid adsorbents – Aluminas, silicas, zeolites, molecular sieves.
- Polymeric compounds – Tenax, chromosorbs and poropaks. These solid polymers can act as extended liquids at higher column temperatures.
- Liquid phases – Common liquid phases include various hydrocarbons, polyglycols, polyethers, polyesters and polysiloxanes. The most widely used are the polysiloxanes.

These are chemically modified and blended to provide a range of stationary phases of different polarities.

Typical applications include:

- Solid adsorbents – Inert gases, freons, very volatile C_1 to C_{10} hydrocarbons.
- Porous polymers – Historically used for packed columns. Good low bleed characteristics. General all round use from hydrocarbons to acids.
- Liquid phases – The most commonly used stationary phases with general applicability to most analytes and matrices. Chemically bonded liquid phases are used in applications involving mass spectrometry and high temperature work.

Figure 10 shows some of the common liquids used, arranged in order of polarity.

4.2 Column Description

Figure 11 shows the information that needs to be specified when a column is purchased.

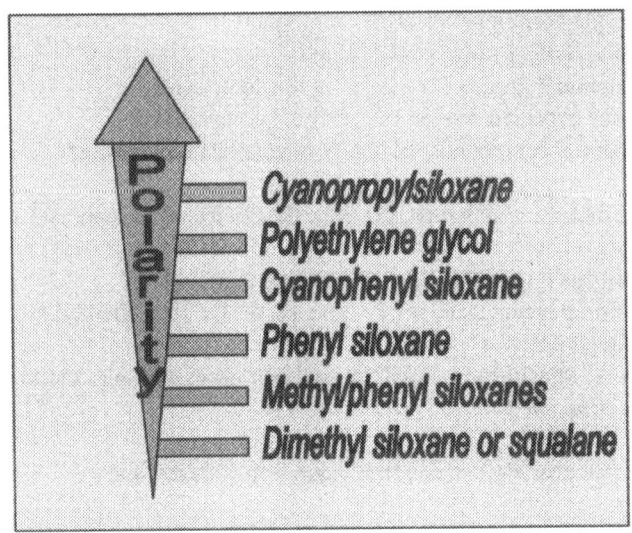

Figure 10 *Common liquid phases arranged in order of polarity*

4.3 Column Selection

For the inexperienced analyst involved in GC method development work on a new or unfamiliar sample, selecting the most suitable column from the wide choice available can be an onerous task. One basic approach to this problem is to marry the particular properties of the sample with those of the capillary column.

The important questions to ask about the sample are:

- What are the sample components and matrix?
- How many compounds are expected in the sample?

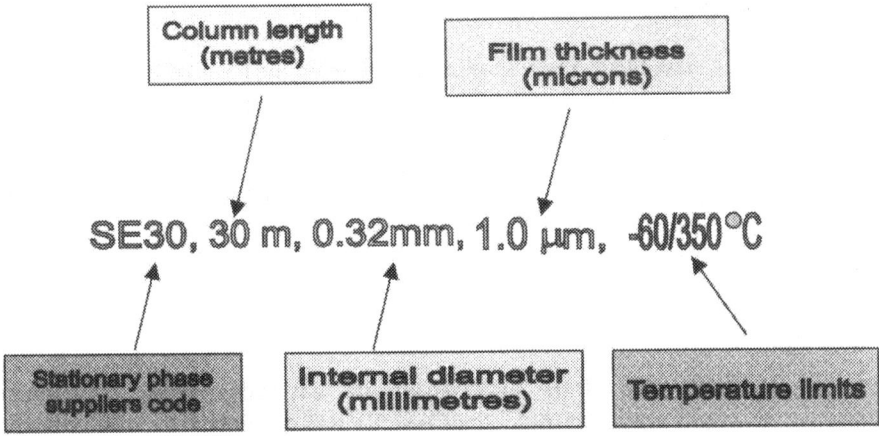

Figure 11 *Column specification*

- What are the boiling points of the compounds?
- What is the boiling range of the sample?
- What functional groups are present in the compounds?
- What is the expected concentration of the compounds?
- Are any of the compounds thermally or chemically unstable?

Additional factors, which may influence the choice of column, are:

- Is the analysis required to be qualitative or quantitative?
- Do the components need to be unequivocally identified, *e.g.* using a mass spectrometer detector?
- Are special injection or sampling methods necessary, *e.g.* thermal desorption, derivatisation, *etc.*?

The important characteristics of the column to consider are:

- 'Like separates like' – for example the most suitable stationary phase for the separation of polar components (of similar volatility) is likely to be a polar material. This rule is less easy to apply when the volatility of the components differs as well as their polarity.
- When using a non-polar column, components will elute in the same order as their boiling points.
- A standard non-polar column (30 m, 0.25 mm, 0.25 μm capillary column) should be able to resolve components having a 2 °C difference in boiling point.
- The separating power of a column increases with column length.
- The separating power increases with film thickness of stationary phase.
- The wider bore columns can tolerate higher loadings of sample and normally lead to better detection limits for the sample components.
- For high temperature work or when using mass spectrometer detectors, high temperature stability and low bleed are important characteristics for the column stationary phase.

Although it is possible to select a suitable column on the basis of these properties it is far more sensible and pragmatic for the analyst to search for information on analyses of the same (or similar) compounds and matrices, *e.g.* in journals, books, application notes *etc.*

Published reference chromatograms are far more informative than lists of properties, sets of parameters or data lists, as they help the analyst to visually assess the performance of a column.

4.4 Tips

Suppliers provide most of the information that the analyst needs for column selection. Columns are listed in terms of polarity and comparability with columns from other suppliers. Information is usually provided on the important selection parameters, temperature stability, column length, bore and film thickness.

- The catalogues often contain application chromatograms for each column indicating the types of substances separated or the analytical methods to which the column is best suited, *e.g.* numbered EPA (US Environmental Protection Agency) methods. This service can also be supplemented by consulting applications scientists and applications databases provided by column suppliers and gas chromatography manufacturers.
- The life of the column will be enhanced the more effort the analyst takes to produce cleaner extracts and/or samples for injection. Ideally, polar, involatile, matrix compounds and particulates should be removed prior to injection.
- Never replace columns when the oven is above ambient temperatures.
- Short lengths of capillary column positioned before and after the main column can protect the main column from the high injector/detector temperatures.
- Always seal the ends of the capillary column (using a spare septum) before storage.
- Most column contamination occurs in the first few centimetres of capillary column; this can be cut away to restore column performance.
- Always use a new ferrule when reconnecting a column.

4.5 Changing the Capillary Column[1]

a) Visually inspect all aspects of the system
   - Cool all heated zones and visibly inspect oxygen and moisture traps. Replace if necessary.
   - Visually inspect injector and detector sleeves, replace if broken or dirty.
   - Replace critical injector seals, detector seals and septum.

b) Prepare column for installation
   - Select make-up and detector gas flow rates.
   - Cut approximately 10 cm from each column end.
   - Install a nut and appropriately sized ferrule on both column ends.
   - Cut an additional 10 cm from each column end to remove ferrule shards.
   - Place the capillary column in the oven.

c) Connect column and confirm flow
   - Connect the column to the inlet at the manufacturer's recommended insertion distance.
   - Set the appropriate column head pressure.
   - Set split, vent, septa purge and any other applicable inlet gas flows.
   - Confirm the flow by immersing column outlet in a vial of solvent.
   - Connect the column to the detector at the manufacturer's recommended insertion distance.

d) Check for leaks
   - Check for leaks at the injector and detector using a thermal conductivity leak detector. *Do not use soaps or liquids to check for leaks.*

[1] Taken from RESTEK 2000 supplier catalogue

- Set injector and detector temperatures and turn on detector.
- *Caution – do not exceed the phase's maximum operating temperature!*

e) Set dead time, condition and calibrate
- Inject a non-retained substance to set a proper dead time.
- Check system integrity by making sure that the dead volume peak does not tail.
- Condition the column at its maximum operating temperature.
- *Caution – do not exceed the phase's maximum operating temperature!*
- Inject a non-retained substance to verify the proper linear velocity (dead time).
- Run test mixtures to confirm proper installation and column performance.
- Calibrate instrument and inject samples.
- Standby operation:
  Short term: in GC with carrier gas flow at 100 °C–150 °C.
  Long term: flame seal or cap ends with septa – keep in dark.

5 Column Oven and Temperature Programming

The purpose of the column oven is to maintain the chromatographic column at a pre-set temperature during the analysis. It has two main modes of operation:

Isothermal mode – The column oven is held at a constant temperature for the duration of the sample elution.

Non-isothermal mode – The oven fan heater is programmable allowing the temperature of the column oven to be increased, or ramped, at controllable rates for part of, or the whole of, the elution cycle. Methods using temperature programming may incorporate several isothermal and temperature ramp steps and these are used to achieve good separation, as illustrated in Figure 12.

The chromatograms shown in Figure 12a and Figure 12b illustrate the limitations of isothermal analysis when the temperature is either too high or too low:

Too low – ill defined peaks and excessively long retention times.

Too high – poor resolution and possible thermal decomposition of components.

The third chromatogram Figure 12c shows how these problems can be overcome by temperature programming. The eight components of the mixture are eluted in a reasonable time with adequate resolution. All have similar peak shapes and detectability.

Because the temperature strongly influences the peak retention times and shapes, all temperature settings must be accurate and reproducible.

5.1 Technical

Typical instrument specification:

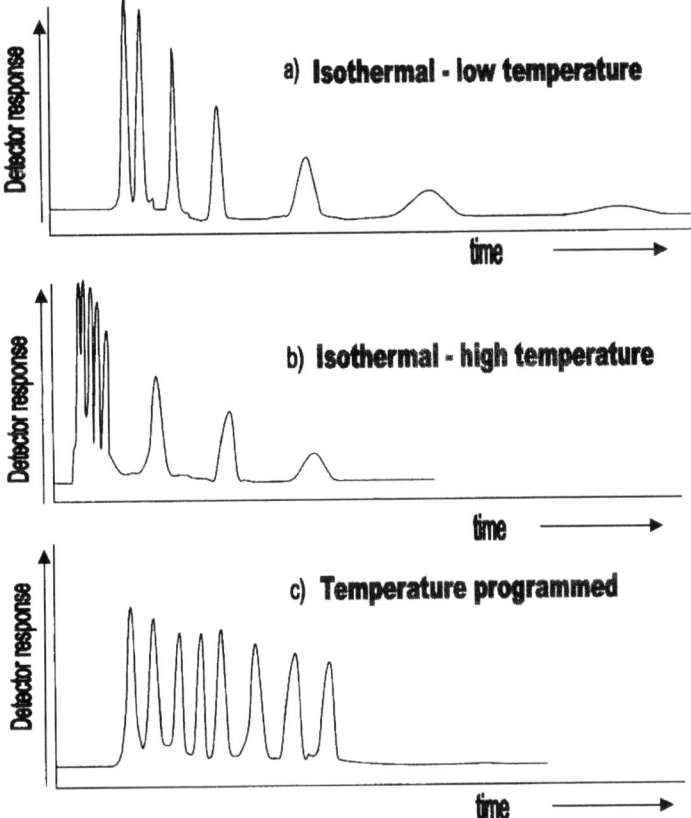

Figure 12 *Typical chromatograms for different column oven temperature modes*

- Convection fan oven.
- Range of temperatures controllable from ambient up to ~500 °C with an accuracy of ±0.5 °C.
- Typical ramp rates in the range 0.5 to 30 °C min^{-1}.
- A ballistic setting – rapid temperature rise ~180 °C min^{-1} for use with on-column or special injection techniques.

5.2 Tips

- Set the optimum flow rate and optimum temperature for the early analyte peaks then use temperature ramp to reduce the elution times and improve the shapes of the later peaks.
- For extracts or matrices containing high boiling components ramp the temperature to a high value at the end of the programme to ensure that all late peaks are eluted before the next sample is injected.
- The maximum column temperature needs to be within the temperature limits of

the stationary phase and should be generally about 25 °C below the detector temperature.

6 Detection

The purpose of the detector is to respond to the sample components as they elute from the end of the chromatographic column, the detector output being appropriate to allow both qualitative and quantitative analysis. For the detector to be useful and fit-for-purpose, it needs to satisfy certain requirements, in terms of sensitivity, selectivity, stability, linear dynamic range, detector cell volume and response time.

6.1 Flame Ionisation Detector (FID)

The flame ionisation detector is the most commonly used detector and key components of its design are illustrated in Figure 13.

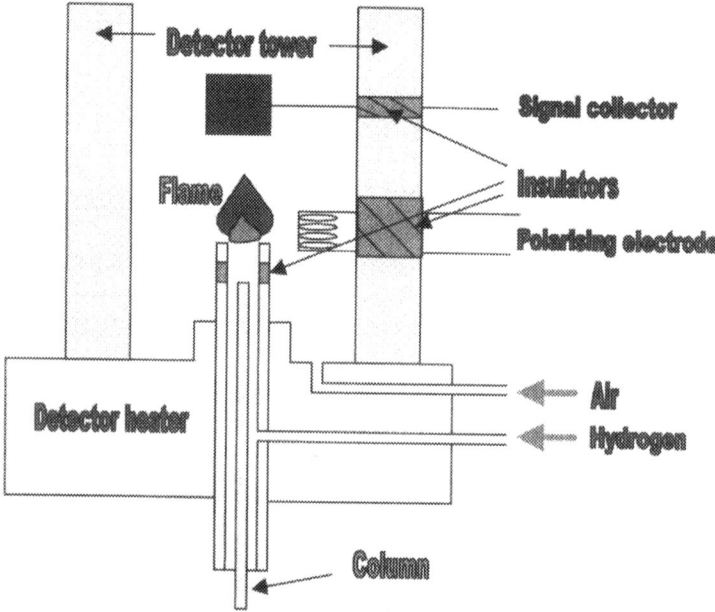

Figure 13 *Schematic of a flame ionisation detector*

6.1.1 Principle of Operation

In the flame ionisation detector the eluant from the column is passed through a hydrogen–air or hydrogen–oxygen flame. When an organic compound containing C–H bonds enters the flame, positively charged ions and electrons are formed as well as the combustion products of carbon dioxide and water. The migration of ions between the electrodes of the detector causes an external current to flow and this allows the organic compounds in the carrier stream to be detected and quantified.

6.1.2 Properties

The properties of the FID are summarised below:

- High sensitivity to most organic compounds.
- Insensitivity to common impurities in the carrier gas, *e.g.* water, carbon dioxide.
- Minimal fluctuations due to changes in flow pressure or temperature.
- Very wide linear dynamic range, up to 8 orders of magnitude.
- The detector has little response to the permanent and inorganic gases (oxidant, fuel and combustion products of the flame) so the background signal from the detector is very small and very stable.

These features have led the FID to be generally regarded as the best 'universal' detector and explain its popularity and extensive use in the field of organic analyses.

6.2 Electron Capture Detector (ECD)

Figure 14 shows the key components of an electron capture detector (ECD).

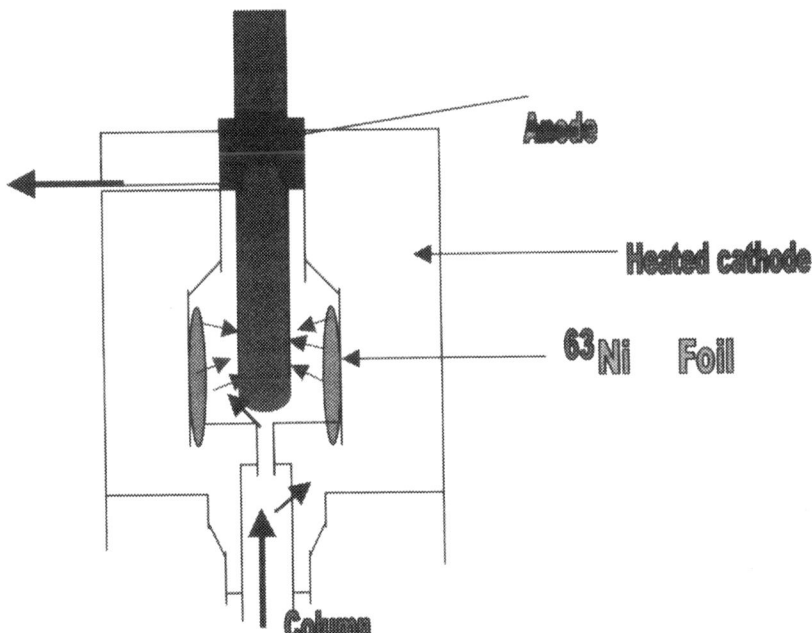

Figure 14 *Electron capture detector*

6.2.1 Principle of Operation

As carrier gas molecules flow past the radioactive nickel source they are ionised to form electrons and positive gas ions. The positive charge on the anode causes the electrons to migrate to the central electrode resulting in a current, often referred to as the 'standing' current, in the external circuit. When an electron-capturing compound enters the detector, electrons are removed and the electron density between the electrodes is temporarily reduced. A corresponding decrease in the external or 'standing' current occurs and the passage of the compound through the detector is registered as a negative peak. In practice the 'standing' current is difficult to maintain using a DC voltage on the anode and this is why a pulsed voltage is often employed.

6.2.2 Properties

- The detector is particularly responsive to halogenated, nitro and organometallic compounds whereas compounds such as hydrocarbons, ethers and esters have almost no electron capturing properties. Because the detector is so specific in its response to particular compounds, as compared with the FID, it is regarded as an example of a 'selective' detector.
- Sensitive to contaminants in the carrier gas, particularly oxygen, and to changes in column temperature.
- Linear dynamic range is poor, usually between 2 to 4 orders of magnitude.
- Short term and long term stability is inferior to the FID detector.

The detector is particularly useful for the analysis of environmentally hazardous chemicals such as volatile halogenated compounds in waters, and samples containing PCBs, dioxins and pesticides. The electron capture detector also finds applications in many other fields where derivatisation is used to facilitate the analysis of a compound. The deliberate formation of halogenated derivatives can lead to enhanced detectability for the compound when GC–ECD is employed in the end analysis.

6.3 Mass Spectrometer Detector (MS)

High-resolution mass spectrometry is a powerful and well-established technique in analytical chemistry, which has been used for over four decades. For organic analyses the equipment is often connected to a gas chromatograph or to thermal probe sources. It is only through the development of bench-top instruments based on the ion trap and quadrupole designs that these new, small low-resolution mass spectrometers have become to be considered as simply a detector for the gas chromatograph.

6.3.1 Principle

Figure 15 shows the basic configuration of an ion trap detector. This illustrates many of the processes that are common to all mass spectrometers.

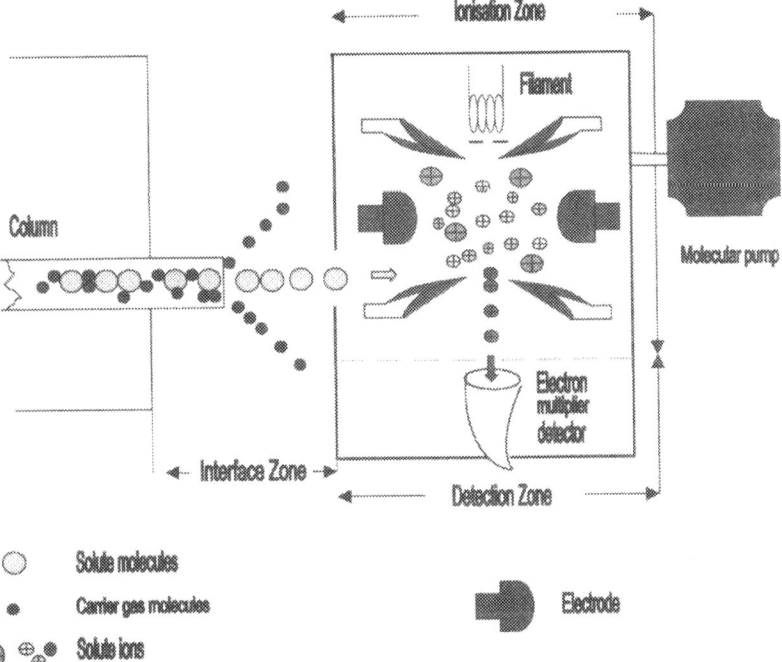

Figure 15 *Schematic of ion trap detector*

6.3.2 Interface

The interface couples the end of the column, which operates at positive gas pressures, to the mass spectrometer, which is operated under high vacuum. The carrier gas molecules are separated by the interface so predominantly only the solute molecules enter the mass spectrometer.

6.3.3 Ionisation and Separation

When the solute molecules enter the trap the filament is switched on and the resulting stream of energetic electrons ionises the solute molecules. The energy of the electrons is controlled at 70 eV, this is sufficient not only to promote ionisation but also causes fragmentation of many of the ions formed. The size and abundance of these fragment ions is characteristic of the solute entering the trap and results in a characteristic mass spectrum of ions.

The ions are initially held in the trap by virtue of a 'storage' radiofrequency voltage applied to the trap electrodes. During the separation stage this field is quickly ramped upwards. The increasing field causes the ions to become unstable and they are ejected progressively from the trap in accordance with their charge to mass ratio, lighter ions being ejected first.

6.3.4 Detection

The funnel shaped detector collects the ions as they leave the trap. As the positive ions collide with the detector, electrons are released in the surface coating compounds. These electrons induce increasingly more electrons to be generated along the length of the detector. This 'cascade' effect produces a highly amplified signal at the detector output. The ionisation, storage and ramp times can all be optimised and controlled by the mass spectrometer's central processor. The total time taken for all these stages (scan time) is normally in tens of milliseconds, this means that within the time it takes for the solute band to elute from the column the scanning process can be repeated many times. A chromatogram is obtained by plotting the total ion signal for each scan against elution time. This total ion chromatogram is almost indistinguishable from a FID trace of the same sample. The mass spectrometer is able to display the ion spectrum for each scan point in the chromatogram. Figure 16a shows the total ion chromatogram for a sample and Figure 16b the mass spectrum for the scan point at the top of the shaded peak shown in Figure 16a.

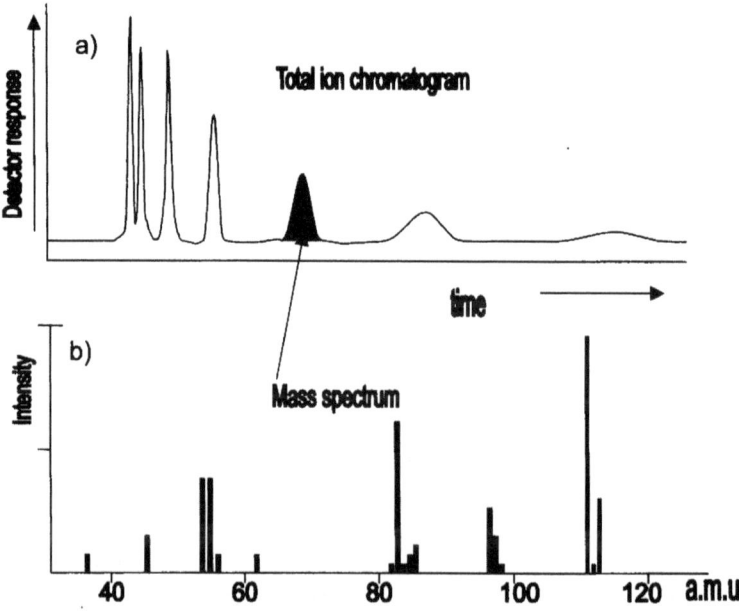

Figure 16 *MS spectral analysis*

6.3.5 Properties

The mass spectrometer offers many features that can be used to facilitate the identification and quantitation of an analyte.

- The detector is a universal type, and is able to detect a greater range of compounds than a FID.
- In the total ion monitoring mode, the detector has similar sensitivity to a FID. Using single ion monitoring (SIM), detection limits can be lowered by several orders of magnitude.
- Linear dynamic range is inferior to the FID. For example, certain ion trap detectors have a modest range of only 3 to 4 orders of magnitude.
- The molecular weight of a compound can be determined using a mass spectrometer detector.
- The compound can identified from its mass spectrum.

6.3.6 Single Ion Monitoring (SIM)

Instead of scanning the whole molecular mass range, normally ~40 to 650 a.m.u. (atomic mass units), the instrument is programmed to detect a single ion mass or very narrow range of ion masses. It is usual to select the most abundant or most characteristic ion mass for the analyte of interest. This can lead to a chromatogram containing only one compound peak. The signal to noise ratio of the compound peak is strongly enhanced by using SIM, leading to a much lower limit of detection for that compound.

6.3.7 Molecular Weight Peak Identification

For many compounds, where ion fragmentation in the mass spectrometer is limited, the molecular ion can be identified in the mass spectrum. This is normally the highest mass ion in the spectrum. Knowledge of the molecular mass and also the chromatographic behaviour of a compound can often lead to its identification, certainly the number of possible candidate empirical formulae is greatly reduced. Bench top instruments normally have unit mass resolution over the whole molecular weight range, which means that the molecular weight of larger compounds can be fixed more accurately than smaller compounds. For compounds where the molecular ion has been lost from the spectrum, much softer ionisation techniques need to be employed to facilitate molecular weight determination, *e.g.* chemical ionisation.

6.3.8 Mass Spectrum

The mass spectrum (ion fragmentation spectrum) for each chromatographic peak can be compared against reference spectra in order to identify a compound. Even the simpler bench top mass spectrometers are supplied with reference libraries, commonly containing the reference spectra of over 40 000 compounds. Libraries can be searched and the best matches to the unknown can be found. A scoring system is used to indicate the closeness of each match with the unknown. Experienced analysts can also interpret the fragmentation patterns to identify the structural units and functional groups present. If this information is combined with data from other spectroscopic techniques it is possible for the analyst to build a complete molecular structure and to identify the compound.

7 Data Handling

7.1 Recording, Manipulating and Reporting Data

The data handling capability of modern equipment has been revolutionised with development of powerful dedicated and/or networked PCs. These systems now automatically acquire, process and report data according to a set of instructions from the analyst, and pre-stored GC methods and auto-sampler sequence files.

The diagram in Figure 17 shows the steps in the data handling process.

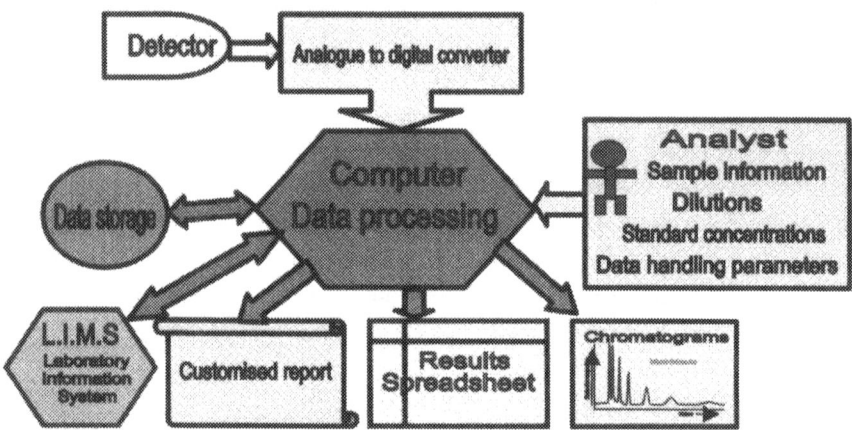

Figure 17 *Schematic of a data handling system*

The analogue signal from the detector is converted to a digital signal by the ADC unit and the digital equivalent of the chromatogram is stored as a raw data file. The analyst enters batch information such as sample size, dilution factors, standard concentrations and data processing parameters into the computer. The processing software identifies component peaks and determines peak areas or heights. The information from samples and standards is used to calculate analyte concentrations and the results are reported in the most appropriate format.

The following example (Figure 18) shows the importance of employing the most appropriate data processing parameters. In each of the chromatograms shown in Figure 18, the analyte peak is shown as the shaded area and the matrix component is shown as the unshaded peak. It should be apparent that the three different integration methods would lead to different peak areas and different analyte concentrations in the sample. However sophisticated the processing software is, peak deconvolution always relies on certain mathematical assumptions and approximations, which can introduce errors into the final results.

The analyst needs to be fully conversant with the operation of the data handling software and hardware to be able to ensure that results are calculated correctly and that reports are tailored to laboratory and customer requirements.

With the most up-to-date software a data log tracks the important instrument parameters during the batch analysis and also registers any alterations to the

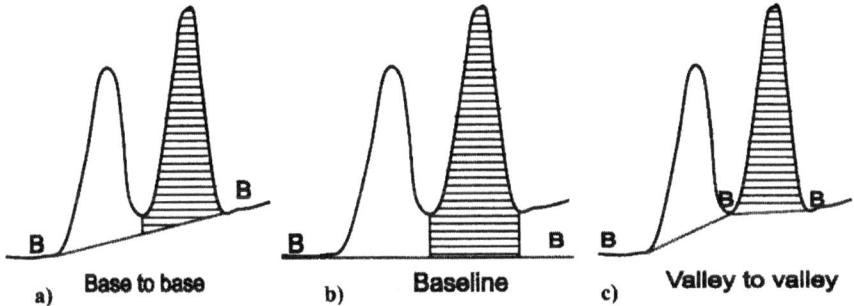

a) **Base to base**   b) **Baseline**   c) **Valley to valley**

Figure 18 *Methods of integrating areas*

methodology introduced by the analyst. If this facility is not provided analysts should record this information in their workbook or instrument logbook. Printed copies of instrument calibrations, chromatograms, intermediate data as well as final reports should be stored if this information cannot be easily archived electronically.

7.2 Tips

- Calculations based on peak area are generally more accurate than those based on peak height because these allow for changes in retention times and peak shape.
- The analyst should strive for the best chromatographic resolution and should not rely on the data handling software to determine the peak areas of unresolved components.
- Check at least one of the calculations that determine analyte concentration in the sample using a pocket calculator or spreadsheet.
- Check that the results from the data processing software are consistent with valid previous work (*i.e.* not corrupted). Use archived data files and results or QC data sets kept specifically for this purpose.
- Record any changes to data handling parameters and archive original versions of the software.

8 Checks, Calibration and Standards

Nowadays it is commonplace for the analyst to be provided with a standard operating procedure (SOP) that details the full analytical method to be used for a particular analyte and matrix. The chromatographic method is also likely to have been pre-developed and often needs only to be called up from the computer methods database. Despite these advantages the analyst is still advised to establish that the components of the gas chromatograph are working correctly and have been properly set up and calibrated prior to the analysis.

8.1 Detector Checks and Calibration

With detectors such as the electron capture and mass spectrometer it is crucial to check that the carrier gas is free from impurities and that there are no air leaks in the equipment. Leaks are indicated by the magnitude of the standing current of the ECD detector whereas mass spectrometer detectors have routines that monitor and determine the amounts of oxygen and water in the carrier and make-up gases. A mass spectrometer has to be separately tuned and calibrated before it is ready to be used as a detector. Normally a pure calibration compound is housed in the instrument for this purpose. Tuning allows the detector parameters to be optimised and the ion masses to be calibrated at numerous points over the complete mass range ~ 30 to 650 a.m.u.

8.2 Standardisation

8.2.1 Check Standard

Where the column, sampling device, injector or detector has been changed to facilitate a particular analysis, it is often expedient for the analyst to run a simple or single component standard as a rapid check on the new chromatographic set-up. For example, manufacturers use a dilute solution of hexachlorobenzene to check the performance of a GC using an electron capture detector. This approach can highlight problems the main program might take hours to reveal. If the check standard has been routinely monitored and logged, the performance of the system can be readily compared with earlier situations.

8.2.2 Grob or Column Calibration Standard

This multi-component standard solution is often received when a new column is purchased. The supplier also provides a reference chromatogram for the standard run on the column prior to dispatch. This standard solution contains a selection of compounds which help demonstrate the performance of the columns with respect to distinct analyte types: Lewis acids, polar compounds, acids and bases, hydrogen bonded compounds *etc*. The solution can also be used to determine column resolution, plate numbers or to monitor the deterioration of the column with time.

8.2.3 Method – External Standardisation

External standards, as their name implies, are materials that are analysed alongside the samples. They are used to identify and quantify the analytes being determined. The calibrant standards normally consist of a series of solutions containing known concentrations of the analytes of interest. For each component, the sensitivity factor (response per unit concentration) is calculated from the standard calibration curve. This is then used to determine the concentrations of analytes in the sample. The calibration range of the standards should always encompass the concentrations of analytes expected in the samples.

8.2.4 *Method – Internal Standardisation*

The purpose of the internal standard is to overcome any potential problems with the GC analysis, particularly those associated with the injection technique, *e.g.* inconsistencies in the volume of sample injected and in the thermal characteristics of the injector.

The detector response factor (R_{factor}) for the analyte compared to that for an identical concentration of internal standard has to be determined. A known amount (volume and concentration) of internal standard is added to a reference standard to produce a solution with equal concentrations of analyte and internal standard. This solution is injected on to the gas chromatograph, and a typical chromatogram is shown in Figure 19.

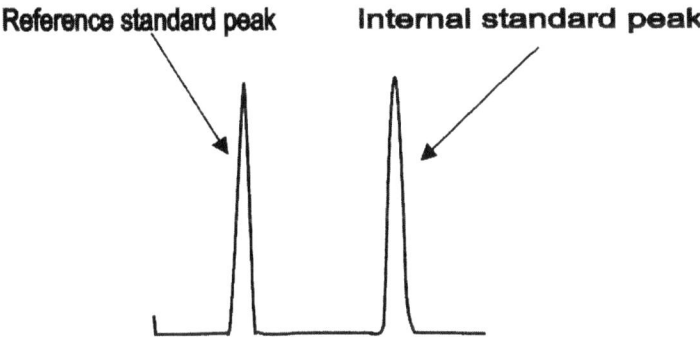

Figure 19 *Reference standard solution chromatogram*

The peak areas are measured and the detector response factor calculated (reference standard peak area/internal standard peak area).

For a sample, a known amount (*i.e.* volume and concentration) of the internal standard is added prior to analysis and injection on to the GC column (Figure 20).

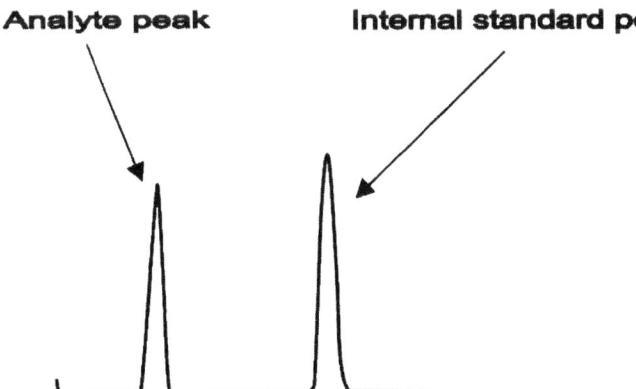

Figure 20 *Sample chromatogram*

The amount of analyte in the sample (Q_{sample}) is then calculated using the following expression:

$$Q_{sample} = \frac{Q_{istandard} \times A_{sample}}{A_{istandard} \times R_{factor}} \tag{6}$$

$Q_{istandard}$ amount of internal standard added to the sample
$A_{istandard}$ internal standard peak area in the sample chromatogram
A_{sample} analyte peak area in the sample chromatogram
R_{factor} detector response factor

This technique is useful for simpler analyses, when the following criteria are satisfied:

- The areas of the peaks for the analyte(s) and internal standard in the sample and reference chromatograms are of similar magnitude.
- The analytes have similar thermal properties and detector responses.
- A detector having a wide linear dynamic range is employed, *e.g.* the FID.

With more complex analyses these criteria are unlikely to be met. In these instances better accuracy and precision (in the end analysis) is gained by using a comprehensive range of external standards (multi-component standard solutions covering a wide range of concentrations) to facilitate the quantitation of the sample analytes. It can still be advantageous to add the internal standard to all the sample and standard solutions (*i.e.* the same amount to each solution) because its peak area in the resultant chromatograms can be used to correct for differences in the injection volume for each solution injected. The internal standard is, in this case, referred to as a 'syringe standard' or 'syringe spike'.

Internal standards can also be added to the sample at the beginning of the analytical procedure, *i.e.* prior to the extraction, clean-up, separation, derivatisation or concentration stages. This is a powerful technique which can, if the internal standard and analytes are well matched, improve the accuracy of the analysis by compensating for any losses of the analytes that occur during the sample preparation procedure. The difficulty for the analyst is in finding an appropriate internal standard(s), which has close physical and chemical properties to all the analytes being determined.

One form of standardisation, which is becoming more and more important, is based on the use of isotope-labelled internal standards. This method requires the use of the mass spectrometer detector, which is able to resolve the slight mass differences in the characteristic ions between unlabelled, 'native', compounds and isotope-labelled compounds. The retention times in most instances are the same.

Known concentrations of labelled compounds are added at an early stage of sample preparation. These labelled homologs are detected alongside the unlabelled analytes in the final extract. Because the chemical and physical properties of the labelled and unlabelled analytes are almost identical, losses of the unlabelled compounds occurring during the extraction, cleanup, concentration stages of preparation and injection stage of the analysis are exactly mirrored by losses in the labelled internal standards. Armed with the response ratio (ratio of the peak response from labelled and unlabelled standards), and the known concentration of unlabelled and labelled compounds in the standards, it is possible to determine, using the measured peak areas, the concentration of the unlabelled analytes in the samples.

8.2.5 Quality Control Standards (QCs)

These materials can be regarded as samples, which contain known concentrations of the analytes being determined. They are analysed periodically with each batch of samples and calibration standards and serve as a check on the accuracy and precision of the analyses. Sometimes QCs are simply standard solutions, which are used to check on the GC end analysis. In other cases, where reference materials are available, the QC can be very similar to the initial sample. In these instances the QC checks the quality of the whole analytical method from extraction to final measurement.

9 Problem-solving

Gas chromatography is a highly sophisticated technique involving many automatic processes, which are often beyond the immediate control of the analyst.

Despite the complexity and refinement of modern equipment, it is reassuring for the analyst to note that the majority of the day to day problems he or she will face will have been encountered on countless occasions by other chromatographers.

Users of the technique are fortunate in the fact that the chromatogram provides a pictorial record of the analysis and this is often very useful for highlighting any problems occurring in the analysis.

9.1 Troubleshooting

The following sections show many of the common problems encountered in gas chromatography and offer suggestions why these occur and how they can be remedied.

9.1.1 Unusual Peak Shapes

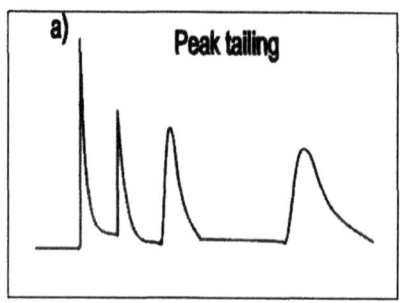
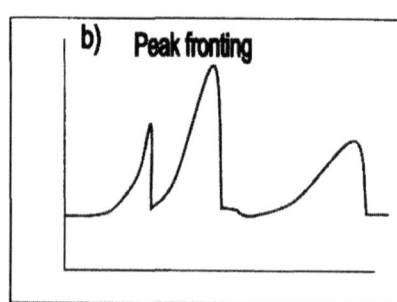
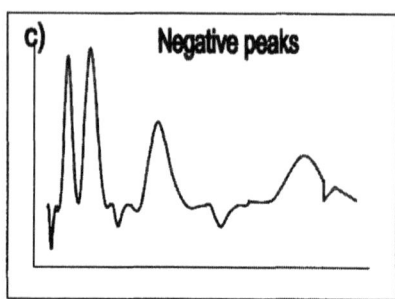
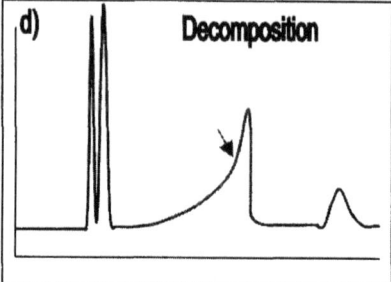

Figure 21 *Unusual shaped peaks*

Table 1 *Unusual peak shapes*

Cause	Cure
a) *Peak tailing* – Adsorption effects in injector or on column.	Clean or replace injector components and column.
Injector or detector temperature too low.	Set injector or detector to correct temperature.
Column not properly installed.	Re-install column.
b) *Peak fronting* – Injector or column overloaded.	Reduce sample size or employ a column of wider bore or higher loading of stationary phase.
c) *Negative peaks* – Dirty EC detector.	Remove anode and clean with solvent, alternatively condition detector at high temperature.
d) *Decomposition* – Component is decomposing in injector or on column.	Check the temperatures of the injector zone and column oven are not too high. Change column to allow lower temperature chromatography.

9.1.2 Additional Peaks

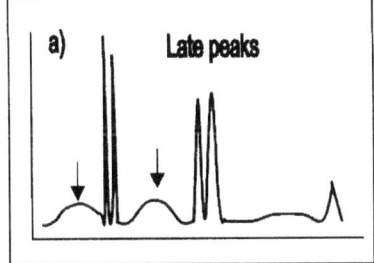
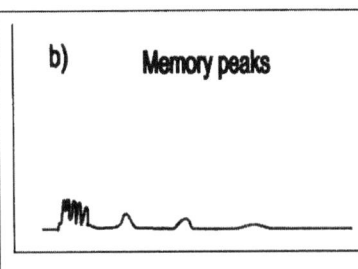
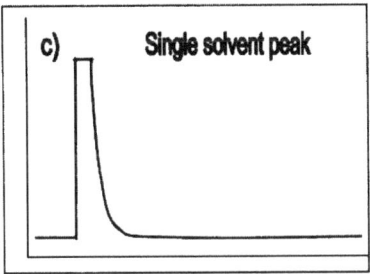

Figure 22 *Additional or missing peaks*

Table 2 *Additional or missing peaks*

Cause	Cure
a) Late peaks – peaks from previous samples still eluting – very broad peaks.	Ramp column to high temperature at end of elution and allow time to clear column.
b) Memory peaks – residual amounts of previous samples are being reintroduced on to the column.	Reduce sample concentrations, increase number of syringe washes, and clean and replace injector components regularly.
c) Single solvent peak – Column unable to retain sample components.	Replace column with one that is more compatible with sample components.

9.1.3 Baseline Problems

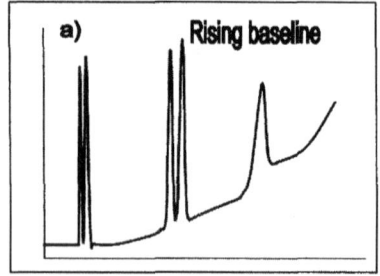
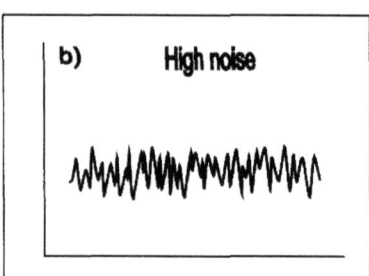
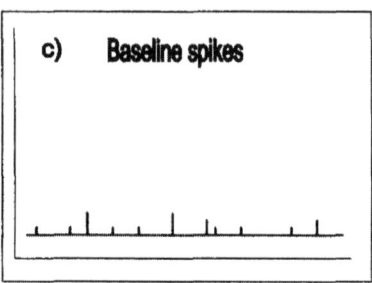

Figure 23 *Baseline problems*

Table 3 *Baseline problems*

Cause	Cure
a) Rising baseline – column bleed exacerbated by temperature program.	Replace column by a polymer type or bonded stationary phase.
b) High noise – Detector parameters incorrectly set or dirty detector.	Reset correct parameters and clean detector if necessary.
c) Baseline spikes – detector dirty or stationary phase particulates emerging from column.	Clean detector and make sure there is a particle trap at the end of the column.

9.1.4 Clipped Peaks

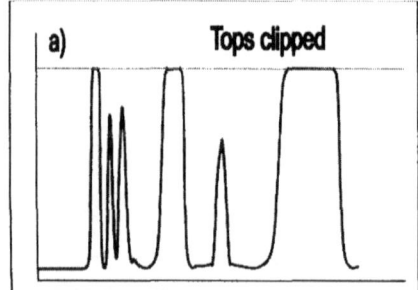
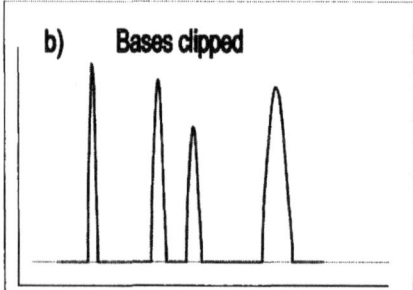

Figure 24 *Clipped peaks*

Table 4 *Clipped peaks*

Cause	Cure
a) *Tops clipped* – Samples too concentrated. Linear range of detector amplifier exceeded or too sensitive detector setting.	Dilute the samples to bring the peaks on scale or use less sensitive detector setting.
b) *Bases clipped* – Integrator zero set too low. Detector drifted below zero.	Reset zero.

10 GC Experiments

To perform these experiments it will be necessary to purchase a standard solution from Thames RESTEK UK Ltd, Fairacres Industrial Centre, Dedworth Road, Windsor, Berks SL4 4LE.

The standard is a 1 mL ampoule of a 'GRO' mix, cat. no. 30069.

The standard solution contains nine compounds at 1000 µg mL^{-1} in methanol.

The compound identities and order of elution are:

1   methanol
2   3-methylpentane
3   2,2,4-trimethylpentane
4   benzene
5   toluene
6   ethylbenzene
7   *m*-xylene
8   *o*-xylene
9   1,2,4-trimethylbenzene
10 naphthalene

Set up the following chromatographic conditions:

- Medium polarity stationary phase equivalent to DB210 or DBwax
- 50–100 metre, 0.53 mm i.d., 3.0 µm
- Split injection
- Injector temperature 250 °C
- Injection volume 0.1 µL
- Detector temperature 250 °C
- Detector – FID
- Carrier gas – helium
- Gas velocity ~ 40 cm s^{-1}

It is appreciated that guide users will have different GC equipment and will almost certainly wish to employ different chromatographic columns for these experiments. The experiments are designed to allow for this and in practice the user may change almost all the conditions detailed above, *e.g.* use packed columns, columns of different length, loading, polarity or use alternative detectors.

10.1 Experiment 1 – Late Peaks

Run the sample at several isothermal temperatures and choose a temperature that gives a chromatogram where all component peaks are eluted in a reasonable time, *e.g.* 20–30 minutes.

The following information will help the user to identify the compound peaks and be sure that all peaks have eluted.

The ethylbenzene, *m*-xylene and *o*-xylene will elute as a tightly grouped set of peaks, the first two probably only partially resolved. Once this group is identified the last eluting peak, naphthalene, will be seen as two peaks later in the chromatogram.

Choose any time value *t* between the elution of *o*-xylene and 1,2,4 trimethylbenzene.

Inject the standard on to the column and at time *t* re-inject the sample again. Monitor and record the second chromatogram for the whole elution period.

The second chromatogram should contain all component peaks expected and also two extra peaks. These are late peaks from the previous injection. They can be readily identified as such because they will be much broader than peaks eluting near to them.

This experiment emphasises the need for the analyst to ensure all components from a sample or extract are eluted before the next sample is injected on to the gas chromatograph.

10.2 Experiment 2 – Resolution

This experiment only involves the separation of the ethylbenzene, *m*-xylene and *o*-xylene peaks.

After these peaks have eluted the oven temperature should be ramped up to a high temperature (200 °C) to ensure that the later compounds are removed from the column before the next injection is made.

a) Change the isothermal temperature, record the chromatogram, and observe the separation of the three selected peaks. Repeat at other isothermal temperatures.
b) Change the carrier gas flow rate (gas velocity) and again observe the separation of the peaks. Repeat again at different flow settings. Note that if the flow rate has to be altered manually (*i.e.* not through the computer) the new flow setting should be calculated from the measurement of dead time, *c.f.* Figure 2.

Tip: Only change one parameter at a time.

These experiments should demonstrate that peak resolution is improved principally by decreasing the temperature and that small improvements are achieved by lowering the gas flow rate.

10.3 Experiment 3 – Change of Carrier Gas

Set the column to low temperature (40–80 °C) and the gas velocity back to 40 cm s^{-1}.

Inject the sample and record the separation of the three peaks.

Change the carrier gas from helium to nitrogen and repeat the experiment.

Note: if this gas change is not practicable or if the quality of the nitrogen gas cannot be guaranteed, do not attempt this experiment. Also remember to protect the column; all heated zones should be left to cool to ambient temperature before the gas change is made.

A decrease in separation should be observed with nitrogen because the 40 cm s^{-1} gas velocity is not an optimum setting for good column performance, *c.f.* Figure 8.

10.4 Experiment 4 – Temperature Programming

Although the best resolution is achieved at low temperatures, this unfortunately leads to very long retention times for the 1,2,4-trimethylbenzene and naphthalene components. Remember that flat broader peaks have poorer detectability than sharp peaks.

a) Try several different temperature programs, containing one or more isothermal and ramp stages, to improve peak shape and detectability of all components.

Clue: Start the program at 40 °C. All peaks in the final optimised chromatogram should have similar peak heights when the optimum conditions are established.

11 A Guide to Finding Information

Peer contacts

Training and advice can be sought from more experienced chromatographers within the same organisation or from outside contacts

Books

There is a plethora of information available, from simple introductory booklets and learning guides to books on the advanced uses of chromatography.

Principles and Practice of Chromatography, B. Ravindrath, 1989, Ellis Horwood Ltd.
Gas Chromatography – Analytical Chemistry by Open Learning, Ian A. Fowlis, 1998, John Wiley & Sons.
Introduction to Analytical Gas Chromatography, Raymond P. W. Scott, 1998, Marcel Dekker Inc.

GC Manufacturers and Suppliers

Magazines often include summary articles and new product information.

Catalogues have product information and application chromatograms.

Applications databases will be cross-referenced to analytes, matrices, columns and specific methods.

Application scientists associated with suppliers and manufacturers.

Troubleshooting guides included in catalogues such as the *Agilent Catalogue 2002/3*.

Courses & Seminars

Seminars are often held on new equipment and chemical analysis applications of GC.

Most manufacturers organise courses on theoretical and practical aspects of GC.

Literature Searches

Journals and reports such as:

Journal of Chromatographic Science
Journal of Chromatography
Journal of High Resolution Chromatography
Journal of LCGC (magazine)

Internet

Large and rapidly expanding source of information.

http://www.aldrich.com/TheReporter
http://www.lcgcmag.com
http://www.agilent.com/chem

Lightning Source UK Ltd.
Milton Keynes UK
UKHW020627010421
381362UK00004B/66